AN ILLUSIONARY MYTH

Omar Lopez, Msc.D.

AN ILLUSIONARY MYTH

Cover Design by

www.srwalkerdesigns.com

ZADKIEL PUBLISHING

A ZADKIEL PUBLISHING PAPERBACK

© Copyright 2022
Omar Lopez, Msc.D.

The right of Omar Lopez to be identified as author and channel of this work has been asserted by him in accordance with the Copyright, Designs and Patents Act 1988.

All Rights Reserved

No reproduction, copy or transmission of the publication may be made without written permission.

No paragraph of this publication may be reproduced, copied or transmitted save with the written permission of the publisher, or in accordance with the provisions of the Copyright Act 1956 (as amended).

Any person who does any unauthorised act in relation to this publication may be liable to criminal prosecution and civil claims for damages.

ISBN: 978 1 78695 810 5

Zadkiel Publishing
An Imprint of Fiction4All
www.fiction4all.com

This Edition
Published 2022

An Illusionary Myth

Omar Lopez, Msc.D.

Terms and Conditions
LEGAL NOTICE

The Author has strived to be as accurate and complete as possible in the creation of this report, even though he does not warrant or represent at any time that the contents within are accurate due to the rapidly changing nature of the Internet.

While all attempts have been made to verify information provided in this publication, the Publisher assumes no responsibility for errors, omissions, or contrary interpretation of the subject matter herein. Any perceived slights of specific persons, peoples, or organizations are unintentional.

In practical advice books, like anything else in life, there are no guarantees of income made. Readers are cautioned to rely on their judgment about their circumstances to act accordingly.

This book is not intended for use as a source of legal business, accounting, or health advice.

TABLE OF CONTENTS

CHAPTER ONE
Are We Living in a Timeless Universe?

CHAPTER TWO
What is the One Electron Theory?

CHAPTER THREE
What is Space-Time?

CHAPTER FOUR
Theory of Biocentrism

CHAPTER FIVE
Is Death an Illusion?

CHAPTER SIX
The Akashic Records

CHAPTER SEVEN
Are we living in a simulation?

CHAPTER EIGHT
The Multiverse Theory

GLOSSARY

ACKNOWLEDGMENTS

Throughout the process of composing this book, Dr. Norman W. Wilson was incredibly helpful in providing direction, comments, and recommendations. I would like to use this opportunity to convey my deepest thanks to him. The fantastic job that Stephen R. Walker from S.R. Walker Designs did designing the book cover warrants special recognition and gratitude on my part. In conclusion, I would like to express my gratitude to www.pixabay.com, www.canva.com and its contributors for granting me permission to use their images and graphics.

DEDICATION

This book is dedicated to the souls that are looking for something beyond what they can currently see, feel, smell, touch and, taste. Aimed towards individuals who are willing to be receptive to new modern ways of thinking as well as the findings of modern scientific research.

If we're distracted from the continual flow of perfect mind that we're in, suddenly everything configures, everything solidifies. Suddenly a shape appears out of flux, a world appears, karmas appear, pasts, futures, presents, time structures, ying and yang appear.

(Frederick Lenz)

CHAPTER ONE

Are We Living in a Timeless Universe?

Is time only a concept, a seemingly functional aspect of subjective experience, or a subjective metric of an unreal entity? Is time only a construct of our brains, or is it a material thing that we ought to manage with care?

The inclusion of time in the equations of physics has been questioned by certain scientists, including quantum gravitational researcher Carlo Rovelli and physicist Julian Barbour, who argue that it should be eliminated.

In any case, time is presented to us as a valuable resource, an objective fact. Our culture instills in us the belief that our lives span a finite period, from birth to death, and that this window of opportunity is a resource to be used sparingly.

"Time is an illusion, a construct made out of human memory. There's no such thing as the past, the present, or the future. It's all happening now."
Blake Crouch (Recursion)

To me, time is more like a tool, something we should pick up, use to get stuff done in the relative world, and then put it back on the shelf when we're done. The clock and calendar are tools for gauging our position in time and space, just like a ruler or other measuring gadget.

Time is useful in that it allows us to schedule appointments, prepare for group activities, take trips, and commemorate special occasions like birthdays, but we have allowed ourselves to be enslaved by it.

We may attest to the accuracy of our concept of time by observing its effects on ourselves. Because of how space behaves close to the gravitational pull of a black hole, physicists believe that time is just an illusion humans create in their minds to make sense of the world in which they live. Time and space at one point were classified into 2 separate things. Today, physicists classify them as Space-Time and are the product of the same things. In the next few chapters, I will go more into detail on the subject.

Science explains quantum theory and non-locality as follows (very roughly): Consider A and B, two particles. You could pair them together, split them up, and send one to either side of the planet or galaxy. Ignore the passage of time as you know it while you stimulate particle A and see particle B's immediate response.

This phenomenon is known as particle entanglement. That is to say, regardless of how far apart they are, the two particles' actions occur simultaneously. I mean, it is a stunning discovery. What in the world is going on here? We need to see some sort of connecting frequency waveform or something to prove that there is a real connection here. This also indicates that particles are interacting at speeds much higher than the speed of light.

Particle Entanglement

At its most fundamental level, quantum studies reveal that our understanding of time as a linear succession of events is off the mark and that there is no mark. Every reference point is merely a convenience that does not exist at the level of fundamental reality.

When we use the term "point" to refer to a specific location in space, we are compressing a vibration field into a more stable state. Pick any location in the universe as your starting point; it might be a speck of dust or anything else. The closer you get to the point, the less distinct it appears to be and the less sense it seems to make, at least from a physical perspective.

This is a serious challenge for scientific inquiry. Not being able to pin anything down precisely creates a convenient excuse for the fuzziness of reality. Here comes Renormalization. Mathematical difficulties arise when trying to define the universe, but scientists have a solution: renormalization. The numbers are rounded both higher and lower.

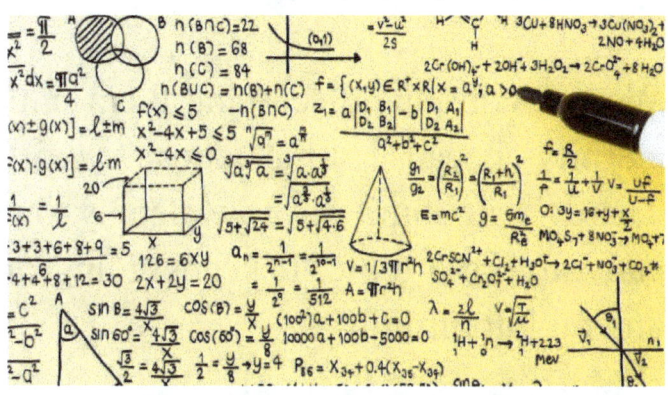

As far as the laws of mathematics refer to reality, they are not certain, and as far as they are certain, they do not refer to reality.
(Albert Einstein)

This inability to give a definitive definition of anything is not just something we observe at the quantum level, but something we witness in our everyday lives as well.

Let's have a conversation about the space you're in, from the ground up: the floor, the carpet, the walls, the paint, the ceiling, the furniture, the appearance, the colors, etc.

We can continue discussing them until eternity because time doesn't exist. There is no limit to how deeply one can explore a topic. It keeps going on and on into infinity.

To begin, let's pretend that from the moment you were born to the moment you woke up, time existed in every part of the universe. Everything that has ever happened, from the Big Bang to the death of the last atom in the universe, is happening right now, in a form of random access universe like a movie film running with no progression from one point to the next or flow of time.

"There is not past, no future; everything flows in an eternal present."
(James Joyce)

This view has more in common with Eternalism, in which the past, present, and future all exist in their own right but there is no method of progression other than that of the observer choosing what to view, and even that idea breaks down without time.

Presentism, in contrast, holds that only the present is real and both the past and the future are unreal. Since the Big Bang, everything in the universe has been in a constant state of transition from one location to another. This chain of events, prescribed by the rules of physics, will continue unabated until all energy in the universe has been equalized and all activity has ceased. According to modern physicists, there would be no you, no me, nothing wouldn't exist if time hadn't existed from the moment of the Big Bang onwards. Without time, nothing could have begun, nothing could have advanced, and nothing would have come into being.

But there's also the potential that everything that ever will happen already has, which raises the question of whether or not we can have the cosmos on our side if we eliminate time. Just as in a dream, everything would appear normal to you, but to someone with a linear time perspective, it would look like a chaotic mess.

It would be like cutting a movie film into pieces, scattering them in the air like shredded film, and then watching each frame at random. Keep in mind that the film can be sliced at different angles creating a different perspective for the observer that is watching it, even though it was the same film. There would be no rationale for birth, childhood, adulthood, or death. I can't imagine how the cosmos would function if there wasn't any time. In reality, even that explanation would be incorrect without time, as it requires time for each word to be spoken and understood, even if they were all spoken at once. Each thought we have is a progression, and if there were no time, it would be impossible to link thoughts or phrases together. The transfer of meaning without time would be an instantaneous tangle of words, letters, sounds, sights, and thought fragments, making it impossible to determine the intended order of those words.

There are instances when we think in terms of time because of the way events play out in our memories. We play things back in our heads as if they were still happening, except the dull, uncomfortable, and humiliating parts are snipped off. Our recall is highly selective. This explains why we tend to have a rosy view of the past and nostalgia.

If we could hop on a photon or other form of a light particle in our regular universe, we'd arrive at a place where time doesn't exist. According to Einstein's theory of relativity, time slows down as you get closer to the speed of light, and since our photon is traveling at the speed of light, time would stop for it, so it wouldn't age at all, even though its trip across the universe took 14 billion years or 1 second. Photons experience no time nor distance in space. As a result, it has not moved through space-time at all, making it a timeless particle.

CHAPTER TWO

What is the One Electron Theory?

What if the cosmos contains just one electron? What if all electrons are the same? The idea that every electron is traveling back and forth in time is put forth in a coherent thought experiment. Here are two questions you've likely never thought about: What unifies all electrons, and why are they unified? If you stop to think about it, there is no reason why every electron in the cosmos would have to have the exact same mass and charge, yet they all do.

A hypothesis developed by physicist John Wheeler in 1940 may help to explain why all electrons are identical. He asserts that every electron in the universe is identical to each other.

There's much intricate physics at play, but to put it briefly: If every electron in the cosmos is the same electron, bouncing backward and forward in space-time, the one-electron hypothesis might be plausible.

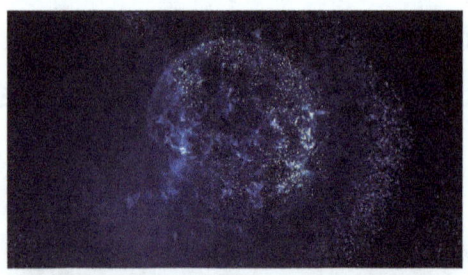

"If you haven't found something strange during the day, it hasn't been much of a day."
(John Archibald Wheeler)

Just as light can make an electron bounce around in space, there might be a way to make an electron bounce backward in space-time. If that's the case, there's one more thing that this theory indicates When electrons move backward in space-time, they become positrons, which are made up of antimatter. Not only are all electrons the same electron, but all positrons are the same electron moving backward. This is another idea, but it hasn't been proven yet.

Since a single electron can go back and forth through time forever, maybe every other particle, from protons and neutrons to neutrinos and other strange and exotic particles, is just one particle going back and forth in time as the Electron Theory explains. If this turns out to be true, that would mean that not only are we all made of the same things, but that each of us is made from just one proton, neutron, and electron going back and forth in space-time.

Tree Of Life

"This oak tree and me, we're made of the same stuff."
(Carl Sagan)

Many find it difficult to accept the idea that time is an illusion generated by the Human mind. I was among those until

I read The Theory of Biocentrism by Robert Lanza.

My contention is that time and the need for it is essentially mental inventions created out of necessity. We observe the earth spinning and the sun setting and rising again. This causes us to want to measure time. There is no such thing as time in the cosmos.

Human beings have been living on our planet for millennia. And some ancient folks were quite intelligent. We have heard that Greeks were outstanding intellectuals, philosophers, and mathematicians that measured and kept records of how the cosmos moved. Then there were the Egyptians and Babylonians, who were both outstanding engineers and built structures in coordination with the position of the stars and planets.

Every beauty which is seen here by persons of perception resembles more than anything else that celestial source from which we all come.
(Michelangelo)

When humanity was confronted with the sequential observations of the sun setting after dawn and rising again in the morning, or the shifting of a star or galaxy from its previous position of observation to a new one, a separate and immediate need arose to measure the time that has passed between two or more sequential events.

Numerous heavenly objects move in unison, whereas others may not. In addition, their movements or occurrences, such as a celestial explosion, are sequential in the cosmos or its aftermath. Numerous events on our planet are also sequential. Changes in the weather, the expansion of buds, their transformation into flowers, fruition, and ripening all occur sequentially. They occur one after another. However, this does not entail the existence of time.

Whenever the human mind witnesses sequential events such as the sunrise or sunset, it is obligated to calculate the epochs that have transpired in between. This acute human necessity to measure these epochs gave rise to the current time measurement systems as we know them today.

When we study history, the passage of so much time since a particular war or conflict can make us sentimental. However, time has not passed on a universal scale. The universe exists only in the present. It is our imaginations that create these perceptions of time passing. The mind generates a need to measure time, and we believe it exists in the cosmos. Our only possession is our moment-to-moment experience of life. "No man ever steps in the same river twice. For it's not the same river and he's not the same man." - Heraclitus

Heraclitus

CHAPTER THREE

What is Space-Time?

A conceptual paradigm that combines the three spatial dimensions with the fourth temporal dimension is called the fabric of space-time. According to the most successful of today's scientific theories, space-time is responsible for explaining both the motion of huge objects throughout the cosmos and the peculiar relativistic phenomena that occur when moving close to the speed of light.

As part of his general theory of relativity, the well-known scientist Albert Einstein contributed to the development of the concept of space-time. Before his groundbreaking work, scientists relied on two distinct models to explain the qualities of physical phenomena. According to NASA, Isaac Newton's laws of physics described the motion of huge objects, and James Clerk Maxwell's electromagnetic models explained the properties of light. Experiments that were carried out during the end of the 19th century gave rise to the hypothesis that light possessed some unique properties. According to the results of the measurements, the speed at which light moved was
constant regardless of the circumstances. And in the year 1898,

the French mathematician and scientist Henri Poincaré proposed the idea that the speed of light might be an impassable barrier. At approximately the same time, several other scholars were contemplating the notion that the dimensions of objects altered based on the speed at which they moved.

In 1905, Einstein published his theory of special relativity, in which he posited that the speed of light was unchangeable. This theory represented the culmination of all of these theories. For this to be true, space and time had to be unified into a single framework that worked together to maintain the same rate of travel for light regardless of who was measuring it.

When compared to a person traveling at a much slower pace, a person in a superfast rocket will see time to be going more slowly and will perceive the lengths of objects to be shorter than they would be if they were traveling at a slower speed.

Because of this, both space and time are considered to be relative, as they are dependent on the rate at which an observer is moving. Both of these things are important, but the speed of light is more basic.

When discussing space-time in today's world, it is common practice to compare it to a sheet of rubber when attempting to explain it. Einstein is also responsible for this discovery. While he was working on his theory of general relativity, Einstein concluded that the force of gravity was caused by curves in the fabric of space-time.

Massive objects, such as the Earth, the sun, or even you, cause space-time to warp and bend as a result of the distortions they induce. These curves, in turn, restrict how everything travels in the universe because objects have to follow paths along this distorted curvature. This curvature is caused by the expansion and contraction of Space-Time. The motion that we experience as a result of gravity is, in reality, motion along the curves and kinks of space-time.

An illustration of a Black Hole curving Space-Time

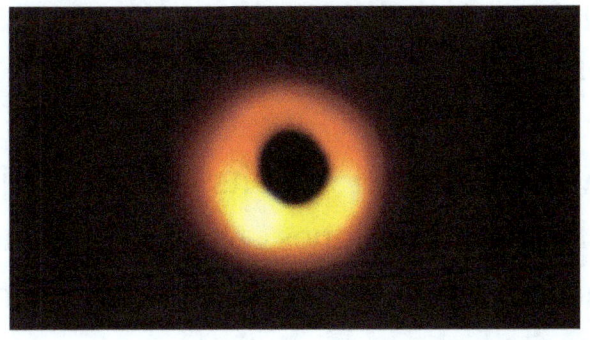

However, the majority of individuals still have trouble wrapping their heads around a lot of this information. Even though we can explain space-time as being comparable to a sheet of rubber, the analogy will inevitably fail at some point. The dimensions of a rubber sheet are two, whereas the dimensions of space and time are four. Not only does the sheet represent warps in space, but it also symbolizes warps in both space and time. Even for physicists, it can be difficult to understand and deal with the complex equations that are utilized to account for everything.

Even with all of its complexities, relativity continues to be the most successful explanation for the various physical phenomena that we are aware of. However, scientists are aware that their models are not accurate because the theory of relativity has not yet been fully reconciled with the theory of quantum mechanics. Quantum mechanics, on the other hand, explains the properties of subatomic particles with an extreme level of precision but does not include the force of gravity.

The foundation of quantum mechanics is the observation that the infinitesimally small components that comprise the universe can be thought of as discretely quantized states. Therefore, photons, which are the particles that makeup light, can be thought of as miniature chunks of light that are delivered in individual packets.

Some theorists have proposed the possibility that space-time itself may also arrive in quantized chunks, which would help to bridge the gap between relativity and quantum physics.

Consequently, if gravity is the deformation of space-time, gravity is a force, and all forces are quantized, space-time itself may consist of discrete units. Perhaps there are fundamental space-time units of an incomprehensibly small size.

The role that space-time plays in physics is one of the most vexing points of contention between general relativity and quantum mechanics. According to quantum mechanics, space-time is merely a backdrop, a stage, a floor, or a container for all the fascinating interactions that comprise the universe's physics. Yes, this stage may bend and warp, and the bending and warping will change the particle pathways, but that's about it. All physics occurs in this space-time background.

Even string theory, the supposed "theory of everything" in which all particles and forces are merely small bits of vibrating strings, implies a background space-time. Therefore, it is a theory of nearly everything.

In general relativity, however, space-time is not the stage upon which the players perform; it is the actor. General relativity does not presume a background; rather, it generates one. General relativity is the language of space-time deformation, and this deformation itself generates the mechanics of gravity.

Consequently, in our attempt to connect quantum physics with gravity, we should possibly take Einstein's hypothesis at face value. If gravity is simply the mechanics of space-time, then we must seek a quantum theory of space-time to develop a quantum theory of gravity. If we can crack this quantization, we will inevitably arrive at a quantum theory of gravity, and the problem will be resolved.

"Space is an illusion, and time as well. There is no such factor as either time or space. We have been blinded by our own cleverness, blinded by false perceptions of those qualities that we term eternity and infinity. "
(Clifford D. Simak)

CHAPTER FOUR

Theory of Biocentrism

Before I explain what the Theory of Biocentrism is, I'd want to present the scientist and genius who originally introduced the concept.

Robert Lanza, a well-known scientist, is the one who first put up the idea of the Theory of Biocentrism. He has worked in fields as diverse as theoretical physics and biology. Prospect magazine named him one of the Top 50 "World Thinkers," and TIME magazine designated him one of the "100 Most Influential People in the World." He worked on the pioneering teams that reported the successful use of pluripotent stem cells in humans and cloned the first human embryo and endangered species.

According to the Theory of Biocentrism, the most fundamental aspects of existence, reality, and the universe are life and biological processes. It explains why life arose in the first place, rather than the universe coming into being first.

The theories of quantum physics are used as a foundation for biocentrism. To build a theory of everything, biocentrism places biology before all other sciences. This is because physics and chemistry are thought to be vital to the study of the cosmos, while biology is considered to be essential to the study of life.

Robert Lanza shows in his explanation of this theory that what humans refer to as space and time are types of animal sensory perception and not exterior physical things. Understanding this simple yet groundbreaking concept presents us with fresh approaches to solving some of the greatest mysteries in contemporary science. Biocentrism provides a novel framework for comprehending everything from the microworld to the forces, constants, and rules that govern the entire cosmos. Biocentrism: How Life and Consciousness are the Keys to Understanding the True Nature of the Universe was first published in 2010 by leading scientist Robert Lanza and astronomer Bob Berman. The book was titled "Biocentrism: How Life and Consciousness are the Keys to Understanding the True Nature of the Universe." As a result of this book, readers from all over the world had a new theory regarding the origin of the cosmos and life, which prompted us to reconsider all we thought we knew. In Western science, the subject of the cosmos has traditionally been dominated by discussions of physics. In his theory known as biocentrism, Robert Lanza shifts the emphasis from physics to biology and asserts that life is not merely an incidental byproduct of physical rules. Instead, it raises the intriguing prospect that life may, at its core, be eternal. This is an encouraging idea.

Beyond Biocentrism: Rethinking Time, Space, Consciousness, and the Illusion of Death is the title of the second book that Robert Lanza and Bob Berman have written on the topic of biocentrism. The authors, take the reader on yet another intellectual and thrilling journey as they re-examine everything we once thought we knew about life, death, the nature of the universe, and the nature of reality itself.

The author acknowledges right off the bat that the way we've been seeing the world up until now is starting to look old and stale in light of recent scientific breakthroughs. This is stated as the basis for the first chapter of the book. Scientists have been able to ascertain that practically all of the universe is made up of dark matter and dark energy, which together make up more than 95% of its total mass. However, scientists must also admit that they do not fully understand what dark matter is and know much less about dark energy. This is something that they must admit openly. In light of this, there is a growing consensus in the scientific community that the cosmos is limitless and exist entirely in our minds.

The purpose of the book *Beyond Biocentrism* is to encourage us to fully accept the repercussions of the most recent scientific discoveries in a variety of fields, such as plant biology, cosmology, quantum entanglement, and even our states of consciousness.

This is intended to be a challenge to us.

When one considers the most recent discoveries in scientific research, it appears as though it is becoming more and more evident that life and consciousness are truly important components of any genuine comprehension of the universe. Because of this, everything that we thought we knew about life, death, and our place in the universe must be reevaluated in light of this new information.

CHAPTER FIVE

Is Death an Illusion?

When it comes to the secrets of our universe and human life, death is the largest mystery that people are intrigued about but also ignorant of. Religions have described death as the only reality and the afterlife as the gift we receive based on our behavior on earth.

But none of this has been demonstrated; nobody knows what the future holds. Robert Lanza, an American medical doctor, and scientist proposes in his book Biocentrism that death is truly a portal to an infinite number of universes.

This phenomenon is referred to as biocentrism, a mechanism that results in all physical possibilities. Biocentrism also emphasizes that it was humans who gave birth to the universe as we know it, as opposed to the universe has given birth to us.

Lanza explains that biocentrism asserts that space and time are merely instruments that our minds employ to weave information into a unified experience, similar to a language for consciousness.

He is a firm believer in the many-worlds hypothesis and asserts that death does not exist in these situations since all of these possibilities are occurring simultaneously and the only reason we feel 'alive' is because of the energy operating in our brains. He claims that the only thing at play here is the energy, and according to the principles of physics, energy can neither be generated nor destroyed; rather, it merely transitions from one condition to another. He claims that this is the case. When we pass away, there is a pause in the orderly progression of the periods and locations we have experienced. Moreover, he contends that our linear conception of time has no bearing on the natural world. In case you were under the impression that he was some crazy person making stuff up, let me assure you that he is knowledgeable. Currently, Dr. Lanza serves as the Head of Astellas Global Regenerative Medicine in addition to serving as the

Chief Scientific Officer of the Astellas Institute for Regenerative Medicine.

His primary area of expertise is stem cell research; however, in recent years, he has also conducted an extensive study in the realms of quantum mechanics and astrophysics. It was in these areas that he developed the notion of biocentrism.

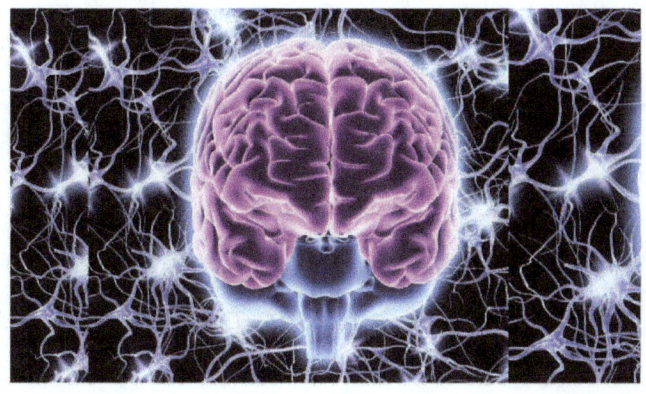

After the passing of his longtime companion, Albert Einstein is quoted as saying "Now that I look back, I see that Besso exited this odd planet a little bit before I did. That is just meaningless. People who think like us are aware of the fact that the division between the past, the present, and the future is an illusion that refuses to give way."

Today, more and more pieces of evidence point to the conclusion that Albert Einstein was correct, and that death is an illusion.

The traditional method of thinking that we adhere to is predicated on the assumption that there is an objective, observer-independent presence in the world. However, a long series of trials demonstrate the exact reverse to be true. We believe that life is nothing more than the interaction of carbon with other molecules; after a little period of existence, we disintegrate into the soil.

We believe in death because we've been told we die. Another obvious reason is that we tend to identify ourselves with our bodies, and we all know that bodies eventually expire. The end of the tale. But according to the emerging theory of everything known as biocentrism, death might not be the final event that we imagine it to be. Surprisingly, some of the most perplexing questions in all of science can be answered once life and consciousness are factored into the equation. For example, it becomes evident why space and time, and even the qualities of matter itself, depend on the observer. This is because the observer's perspective changes everything. It is also made obvious why the laws, forces, and constants of the universe appear to be extraordinarily well-suited for the development of living things.

Any endeavor to comprehend reality will continue to be a dead end until we realize that the cosmos exists entirely within our minds.

Think about how the weather is 'outside': Even though you see a blue sky, the cells in your brain have the potential to be reprogrammed such that the sky appears green or red instead. In point of fact, with just a little bit of work in the field of genetic engineering, we could probably make everything that is red vibrate, make a noise, or even stimulate a desire to have sex in a person, similar to the way that it does in some species of birds. You may believe that it is bright outside, yet the neural circuits in your brain may be altered such that it appears that it is night. You may believe it is warm and muggy, but a tropical frog would experience the same conditions as being chilly and dry. This line of reasoning can be applied to almost every situation. In a nutshell, what you are observing could not be here without your consciousness making it possible.

There is nothing visible through the bone that encircles and protects your brain. Your eyes do not serve as entrances to the outside world. Everything that you are seeing and experiencing right now, including your body, is nothing more than a whirlwind of information that is taking place inside of your head. According to biocentrism, space and time are not the rigid and unfeeling entities that we typically perceive them to be. Move your hand through the air and think about what would be left if everything was taken away. Nothing. The same can be said for the passage of time. Both time and space are merely instruments that are used to piece everything together.

Consider the well-known experiment with the two slits. When researchers observe a particle going through a barrier that has two slits in it, the particle acts like a bullet and either passes through one of the apertures or the other. However, if you are not paying attention, it will behave like a wave and attempt to pass through both slits simultaneously. Therefore, how is it that the behavior of a particle can alter depending on whether or not you are seeing it? The answer is quite straightforward; the nature of reality is a process that includes your consciousness.

Or, take the well-known uncertainty principle developed by Heisenberg. If there is a world out there with particles just bouncing around, then we ought to be able to quantify all of their attributes if this is the case. But you can't. For instance, it is impossible to know both the precise location of a particle and its momentum at the same time. Why, then, should it make a difference to a particle which property you choose to measure? And how is it that two particles that are entangled with one another can be instantly connected on opposite sides of the galaxy as if space and time do not exist? Again, the explanation is quite straightforward: space and time are merely mental constructs that we employ to organize our experience of the world.

CHAPTER SIX

The Akashic Records

According to the late Edgar Casey, the Akashic Records can be thought of as the supercomputer of the universe as well as central storage for all information. The Akashic Records are a collective field of energy that holds the knowledge of everything that has ever existed, as well as everything that will ever exist. It holds the past, present, and future knowledge of all things. And it can be found in all places at the same time. Individual energy fields for each beam can be found contained inside this overarching field; these energy fields are extremely extensive in their own right.

The Akashic Records can be compared to the widespread Wi-Fi networks of today. Although we are unable to see the Wi-Fi internet, we are aware that it is present all around us. It is something that cannot be seen, but if you know how to gain access to it, you will have access to an infinite amount of information. You may locate these hidden secret patterns and codes that are circulating around us pretty much anywhere on the earth that has a connection to the internet.

The Akashic Records can be thought of in the same way. To gain access to this information, all that is required of you is to tune your mind to the appropriate frequency.

Many people talk about or explain the idea in different ways, but the Akashic Records are thought to be the place where all the thoughts, words, and actions of every living thing, good and bad, from all times, past, present, and future, are kept. But people who know about the records say that there is no judgment or implied punishment in them. Instead, the records are just a record of each soul's journey through the infinite, they say.

The Akashic Records are a massive storehouse of information that includes unimaginable levels of wisdom and comprehension. It is a wonderful resource to consult if you have ever wondered why you are the person you are, how you arrived at this point, or what future possibilities you may find yourself experiencing during the course of this lifetime. Because the Akashic records include a history of our souls, anyone can learn how to read them. Everyone has access to this information, but the degree to which they receive it and the amount of information they receive will differ depending on the personal and distinctive senses they possess. Because of their attention to detail, perhaps certain individuals are given more complicated pieces of information. Some individuals get messages through the use of specific colors, while others do it through movement to music, dance, or solving math equations.

Generally speaking, you can gain access to the records by meditating, praying, or setting a goal and asking the cosmos for it. This intention or prayer is a vibrational frequency like a mantra made up of particular sounds that grant you access to the Akashic records. It allows for a great deal more flexibility about the regulations and the process of entering the records. If you are successful in unlocking this guidebook of support, you will have access to it for the remainder of your life. It is the same as carrying around a manual of your past, present, and future life in your rear pocket.

Throughout the course of human history, human beings have engaged in interaction, both consciously and unconsciously, with this body of energy. These encounters have been both planned and unplanned at the same time. This collection of learning has been an ever-present source of spiritual support for everyone, including those in our Western Judeo-Christian tradition. It is referred to in practically every major traditional religion as the Book of Life and the Book of God's Remembrance.

Because of the tremendous quantity of esoteric and occult literature that has been connected to the name Akasha (Sanskrit aka) since the beginning of the nineteenth century, it is quite difficult to pinpoint where the term Akasha (Sanskrit aka) originally came from. Because of this, it is challenging to determine the point at which the original idea stops and the Western embodiment of it begins.

There is a strong possibility that the name Akasha was derived from an early Western linguist's attempt to transliterate the Sanskrit word ka. The essence of the word ka can be rendered in English as either space, atmosphere, or sky. Depending on the context, it might have a variety of additional meanings.

Numerous great teachers such as philosophers, mathematicians, physicists, and other human beings have tapped into this treasure of information and transformed the world as a result.

"My brain is only a receiver in the universe, there is a core from which we obtain knowledge, strength, and inspiration. I haven't penetrated the secrets of this core, but I know that it exists"
(Nikola Tesla)

Nikola Tesla, the man regarded as having the greatest mind of all time, was not only an inventor or, as many people assert, the genuine father of the electric age; he was also a guy who possessed secrets and riddles that we could not even begin to fathom. Even though he was famous, not as much as he should have been because most of the patents were registered by Edison or Westinghouse, Tesla himself was altruistic, and when he learned of others using his ideas, he probably wished them the best of luck. This is because Edison or Westinghouse registered the majority of the patents. He was not motivated monetarily beyond the need to survive, and he desired for his thoughts to be communicated to the entirety of the globe, either by him or by others.

Einstein was once asked what it was like to be the smartest man alive. He replied, "I don't know, you'll have to ask **Nikola Tesla**."

Nikola Tesla

CHAPTER SEVEN

Are we living in a simulation?

What exactly is the Simulation Theory? The idea behind the simulation theory is that we are all merely virtual beings that are merely carrying out our lives in a computer simulation. Therefore, it should not come as a shock that the topic is once again making its way to the forefront when one takes into consideration the rapid pace of technological advancement and evolution.

In what ways are we getting closer to being able to fully program the physical world?

The internet of senses is one of the three main drivers that Ericsson has identified as being most significant to the evolution of the network platform. These key drivers are all connected to the concept of bridging the gap between the physical world and the digital world in some way. The creation of cyber-physical systems brings us one step closer to the production of a digital reality that is believed to be real.

Connected intelligent machines are another important driving factor. The more capable they are of learning on their own and interacting and communicating with one another, particularly with the assistance of AI-to-AI communication and autonomous systems, the more likely it is that we will see them generate hypotheses and reasoned arguments, make recommendations, and take actions. Having said that, do you think that one day intelligent machines will be able to reach the same level of artificial intelligence as humans? In addition, will it be possible for us to develop a simulation that is nested within a reality that is believed to be real?

The game industry arguably comes the closest to creating a simulation that is believable at this point in time. Consumers anticipate that this industry will be the first to offer a combined visual experience of this kind. This might not come as a surprise to you. More than seven out of ten participants feel that by the year 2030, virtual reality game worlds will be indistinguishable from the real world.

Additionally, virtual reality is making available considerably more creative opportunities for the digital portrayal of a physical object or location in the actual world.

Elon Musk does not believe that you are real. But this has nothing to do with him personally; he believes that he, too, does not exist. At the very least, not in the conventional sense of the word "exist." Instead, humans are merely immaterial software constructs that are being simulated by a super alien computer. Musk has suggested that the likelihood that we are in fact living in "base reality," often known as the physical universe, is one in a billion billion billion.

Surprisingly widespread support can be found not only among philosophers but also among many scientists, physicists and medical doctors. Its updated form is derived on an influential work written in 2003 by the Swedish philosopher Nick Bostrom titled "Are We Living in a Computer Simulation?" Assume, he adds, that in the distant future, civilizations who are enormously more technologically advanced than ours will be interested in running "ancestor simulations" of the sentient beings that existed in their far distant galactic past. In such case, the number of simulated minds will one day vastly outnumber those of genuine minds. You should be quite astonished if it turns out that you are one of the few true brains in existence rather than one of the trillions of simulated minds; as a result, you should be very surprised.

The notion that humans are incapable of knowing anything with absolute certainty about the world outside of ourselves has a long and illustrious history in both philosophical skepticism and other traditions. Zhuangzi, a Taoist sage from China, is credited with writing a well-known fable about a man who couldn't decide whether he was a butterfly dreaming of being a man or a man dreaming of being a butterfly. Zhuangzi's story is a classic parable. The French philosopher René Descartes had the idea that he was being controlled by an "evil demon" (or an "evil genius") that was responsible for all of the sensations that he experienced, while the American philosopher Hilary Putnam of the 20th century coined the phrase "brain in a vat" to describe an idea that was conceptually similar. However, according to the simulation hypothesis, you do not have a physical body anyplace. You are nothing more than the product of complex mathematical processes carried out on some enormous super computer.

Our brain simulates reality. So, our everyday experiences are a form of dreaming, which is to say, they are mental models, simulations, not the things they appear to be.
(Stephen LaBerge)

Stephen LaBerge is an American psychophysiologist specializing in the scientific study of lucid dreaming. In 1967 he received his bachelor's degree in mathematics. He began researching lucid dreaming for his Ph.D. in psychophysiology at Stanford University, which he received in 1980.

CHAPTER EIGHT

The Multiverse Theory

According to the Multiverse Theory, the immense region of space known as "The Universe" is not the only thing that exists. Even though the stars and galaxies of our visible universe extend beyond infinity, there are ideas that suggest other universes may exist in parallel dimensions or that the enormous region of our universe is just one of many such cosmic expanses.

If there is a multiverse, the sum of all that exists would consist of several universes that can be viewed as "alternative realities," "parallel universes," multiple timelines, and many-worlds interpretations. Since the presence of other universes has been an arcane concept throughout scientific history, many academics say that it is practically hard to establish or refute multiverse theories using conventional research techniques.

There are still renowned physicists who believe that a multiverse model provides the best explanation for the illogical behavior of quantum particles such as photons and electrons. The Multiverse Theory is a matter of scientific discussion that should be regarded with a grain of salt, however the accompanying guide describes the theory for those who wish to examine it and form their own opinion regarding its plausibility.

How does the theory of the multiverse work? There are numerous approaches to get at a model of multiple universes. Either our universe lives in an area of space, or "bubble," of which there are many, or there could be alternative universes with distinct timelines. There is also the possibility of higher dimensions in which an n-dimensional matrix of universes contains multiple 3-dimensional worlds similar to our own.

Max Tegmark's theory classifies multiverses into four levels, with the higher levels expanding onto the lower levels. The first level of other worlds, according to the theory, are those that merely expand beyond the visible one. The second-level universes of the multiverse exist outside our own "bubble" of galaxies, separated from our universe by enormous voids devoid of galaxies and stars. Before the 1930s, it was believed that the Milky Way was the entire universe, and other galaxies were referred to as "island universes".

Level three universes are likewise three-dimensional, similar to our own, but they branch off from other universes via quantum interactions to form an eternally branching multiverse. Tegmark characterizes the fourth level as an all-encompassing theory of everything that exists, in which case all multiverse theories and other sorts of parallel universes would have to exist within this fourth completely infinite collection of worlds.

The concept of numerous universes extends all the way back to Ancient Greece. The earliest known instance of a doctrine of numerous worlds can be found in the Greek philosophy of Atomism. Despite the fact that the atom was not confirmed by science until many centuries after the time of Ancient Greece, some intellectuals at the time believed that all matter was formed of tiny, indivisible units they called atoms. According to a portion of this hypothesis, collisions between these particles might produce new worlds.

Chrysippus, an ancient Greek philosopher, postulated that the cosmos likely expires and is recreated, which would entail the existence of an endless number of universes in the past and future.

"Time and space are illusions. Everything exists at the same time. We only see what we are tuned to the vibration of to see.

As we change our ideas, we change our vibrations, we start to see a different world, literally. Because we have shifted our consciousness, our focus, to a different version of Earth that exists simultaneously with the version we were on a moment ago. And we are experiencing literally, bit by bit, whether fast or slow, a progression through different versions of Earth.

So the change has to be made within the person. They may for a while still be able to see and observe the version of Earth that they are no longer really strongly connected to, but that will change over time.

It's not that the world they were on changes, it's that what changes is their ability to still perceive that world as opposed to perceiving one that's now more in alignment with the vibrational frequencies they have changed within themselves."

(Darryl Anka)

Is the existence of the multiverse possible? Scientists increasingly regard the notion of higher dimensions and parallel universes as one of the best approaches to explain how quantum theory operates in the field of current physics. However, traditional laboratory procedures are incapable of proving or disproving such theories. Some opponents of the theory cite the "Fermi Paradox," which states that, in a vast multiverse of potentially infinite parallel universes inhabited by advanced lifeforms, it seems mathematically certain that one of them would have visited us by now, but there is no conclusive evidence that this has actually occurred.

Still unknown to physicists is the fate of objects that perish within a black hole. Some scientists hypothesize that our universe could have formed from the opposite side of a black hole in a parallel universe and that a vast number of similar universes have been generated by black holes based on their examination of what happens to stars and other objects that are drawn into them.

The "String Theory" has spawned several multiverse theories to explain the strange behaviors of quantum particles, which include instantaneous teleportation, entanglement with other particles despite vast distances, and simultaneous existence in more than one state or location, a paradox that mathematicians have struggled to validate. M-theory is a popular interpretation of string theory that describes the most fundamental components of existence as one-dimensional loops vibrating at distinct frequencies that define their nature and function.

M-theory and string theory are both attempts to give a single theory that unites the macrophysics of the observable universe with the quantum physics of the micro universe. String theory and M-theory propose additional dimensions in addition to the three dimensions we interact with and experience, namely height, width, and length. The mathematics of quantum theory has been discovered to work better within models of reality with 10 or more spatial dimensions and one or more time dimensions. This has led to more acceptance for additional dimensions.

"A human being is a part of the whole called by us universe, a part limited in time and space. He experiences himself, his thoughts and feeling as something separated from the rest, a kind of optical delusion of his consciousness. This delusion is a kind of prison for us, restricting us to our personal desires and to affection for a few persons nearest to us. Our task must be to free ourselves from this prison by widening our circle of compassion to embrace all living creatures and the whole of nature in its beauty."

(Albert Einstein)

About the Author

Mystery of Metaphysics & Existence was written by Omar Lopez, who is also the author of this book. Although he spent the majority of his life up until recently in Pennsylvania, he was born in Weehawken, New Jersey. When he's not writing, you can find him volunteering as an online course developer.

Since Omar began his career, he has spent the last quarter of a century working in the industry as an electronic technician. As Omar thought for days on end about many topics, including "How does the universe work?", "Are we living in a timeless universe?", Does time even exist or is it a construct created by the human mind to better comprehend the world around us?

Lopez holds a doctorate in Metaphysics from U.L.C. in Modesto, California, and is

Board Certified by The American Association of Drugless Practitioners. Prior to beginning his career as a writer, Lopez pursued and obtained these credentials. After that, in order to change things up a bit, he is going back to school in order to get his second doctorate, this time in Metaphysical Humanistic Science from Thomas Francis University.

GLOSSARY

A

Absolute Magnitude
A scale for measuring the actual brightness of a celestial object without accounting for the distance of the object. Absolute magnitude measures how bright an object would appear if it were exactly 10 parsecs (about 33 light-years) away from Earth. On this scale, the Sun has an absolute magnitude of +4.8 while it has an apparent magnitude of -26.7 because it is so close.

Absolute Zero
The temperature at which the motion of all atoms and molecules stops and no heat is given off. Absolute zero is reached at 0 degrees Kelvin or -273.16 degrees Celsius.

Ablation
A process by where the atmosphere melts away and removes the surface material of an incoming meteorite.

Accretion
The process by where dust and gas accumulated into larger bodies such as stars and planets.

Accretion Disk
A disk of gas that accumulates around a center of gravitational attraction, such as a white dwarf, neutron star, or black hole. As the gas spirals in, it becomes hot and emits light or even X-radiation.

Achondrite
A stone meteorite that contains no chondrules.

Albedo
The reflective property of a non-luminous object. A perfect mirror would have an albedo of 100% while a black hole would have an albedo of 0%.

Albedo Feature
A dark or light marking on the surface of an object that may or may not be a geological or topographical feature.

Altitude
The angular distance of an object above the horizon.

Antimatter
Matter consisting of particles with charges opposite that of ordinary matter. In antimatter, protons have a negative charge while electrons have a positive charge.

Antipodal Point
A point that is on the direct opposite side of a planet.

Apastron
The point of greatest separation of two stars, such as in a binary star system.

Aperture
The size of the opening through which light passes in an optical instrument such as a camera or telescope.

Aphelion
The point in the orbit of a planet or other celestial body where it is farthest from the Sun.

Apogee
The point in the orbit of the Moon or other satellite where it is farthest from the Earth.

Apparent Magnitude
The apparent brightness of an object in the sky as it appears to an observer on Earth. Bright objects have a low apparent magnitude while dim objects will have a higher apparent magnitude.

Asteroid
A small planetary body in orbit around the Sun, larger than a meteoroid but smaller than a planet. Most asteroids can be found in a belt between the orbits of Mars and Jupiter. The orbits of some asteroids take them close to the Sun, which also takes them across the paths of the planets.

Astrochemistry
The branch of science that explores the chemical interactions between dust and gas interspersed between the stars.

Astronomical Unit (AU)
A unit of measure equal to the average distance between the Earth and the Sun, approximately 93 million miles.

Atmosphere
A layer of gases surrounding a planet, moon, or star. The Earth's atmosphere is 120 miles thick and is

composed mainly of nitrogen, oxygen, carbon dioxide, and a few other trace gases.

Aurora
A glow in a planet's ionosphere caused by the interaction between the planet's magnetic field and charged particles from the Sun. This phenomenon is known as the Aurora Borealis in the Earth's northern hemisphere and the Aurora Australis in the Earth's Southern Hemisphere.

Aurora Australis
Also known as the southern lights, this is an atmospheric phenomenon that displays a diffuse glow in the sky in the southern hemisphere. It is caused by charged particles from the Sun as they interact with the Earth's magnetic field. Known as the Aurora Borealis in the northern hemisphere.

Aurora Borealis
Also known as the northern lights, this is an atmospheric phenomenon that displays a diffuse glow in the sky in the northern hemisphere. It is caused by charged particles from the Sun as they interact with the Earth's magnetic field. Known as the Aurora Australis in the southern hemisphere.

Axis
Also known as the poles, this is an imaginary line through the center of rotation of an object.

Azimuth
The angular distance of an object around or parallel to the horizon from a predefined zero point.

B

Bar
A unit of measure of atmospheric pressure. One bar is equal to 0.987 atmospheres, 1.02 kg/cm2, 100 kilopascal, and 14.5 lbs/square inch.

Big Bang
The theory that suggests that the universe was formed from a single point in space during a cataclysmic explosion about 13.7 billion years ago. This is the current accepted theory for the origin of the universe and is supported by measurements of background radiation and the observed expansion of space.

Binary
A system of two stars that revolve around a common center of gravity.

Black Hole
The collapsed core of a massive star. Stars that are very massive will collapse under their own gravity when their fuel is exhausted. The collapse continues until all matter is crushed out of existence into what is known as a singularity. The gravitational pull is so strong that not even light can escape.

Black Moon
A term used to describe an extra new that occurs in a season. It usually refers to the third new moon in a season with four new moons. The term is sometimes used to describe a second new moon in a single month.

Blue Moon
A term used to describe an extra full that occurs in a season. It usually refers to the third full moon in a season with four full moons. Note that a blue moon does not actually appear blue in color. It is merely a coincidence in timing caused by the fact that the lunar month is slightly shorter than a calendar month. More recently, the term has also been used to describe a second full moon in a single month.

Blueshift
A shift in the lines of an object's spectrum toward the blue end. Blueshift indicates that an object is

moving toward the observer. The larger the blueshift, the faster the object is moving.

Bolide
A term used to describe an exceptionally bright meteor. Bolides typically will produce a sonic boom.

C

Caldera
A type of volcanic crater that is extremely large, usually formed by the collapse of a volcanic cone or by a violent volcanic explosion. Crater Lake is one example of a caldera on Earth.

Catena
A series or chain of craters.

Cavus
A hollow, irregular depression.

Celestial Equator
An imaginary line that divides the celestial sphere into a northern and southern hemisphere.

Celestial Poles
The North and South poles of the celestial sphere.

Celestial Sphere
An imaginary sphere around the Earth on which the stars and planets appear to be positioned.

Cepheid Variable
This is a variable star whose light pulsates in a regular cycle. The period of fluctuation is linked to the brightness of the star. Brighter Cepheids will have a longer period.

Chaos
A distinctive area of broken terrain.

Chasma
Another name used to describe a canyon.

Chondrite
A meteorite that contains chondrules.

Chondrule
Small, glassy spheres commonly found in meteorites.

Chromosphere
The part of the Sun's atmosphere just above the surface.

Circumpolar Star
A star that never sets but always stays above the horizon. This depends on the location of the observer. The further South you go the fewer stars will be circumpolar. Polaris, the North Star, is circumpolar in most of the northern hemisphere.

Circumstellar Disk
A torus or ring-shaped accumulation of gas, dust, or other debris in orbit around a star in different phases of its life cycle.

Coma
An area of dust or gas surrounding the nucleus of a comet.

Comet
A gigantic ball of ice and rock that orbit the Sun in a highly eccentric orbit. Some comets have an orbit that brings them close to the Sun where they form a long tail of gas and dust as they are heated by the Sun's rays.

Conjunction
An event that occurs when two or more celestial objects appear close together in the sky.

Constellation
A grouping of stars that make an imaginary picture in the sky.

Corona
The outer part of the Sun's atmosphere. The corona is visible from Earth during a total solar eclipse. It is the bright glow seen in most solar eclipse photos.

Cosmic Ray
Atomic nuclei (mostly protons) that are observed to strike the Earth's atmosphere with extremely high amounts of energy.

Cosmic String
A tube-like configuration of energy that is believed to have existed in the early universe. A cosmic string would have a thickness smaller than a trillionth of an inch but its length would extend from one end of the visible universe to the other.

Cosmogony
The study of celestial systems, including the Solar System, stars, galaxies, and galactic clusters.

Cosmology
A branch of science that deals with studying the origin, structure, and nature of the universe.

Crater
A bowl-shaped depression formed by the impact of an asteroid or meteoroid. Also the depression around the opening of a volcano.

D

Dark Matter
A term used to describe matter in the universe that cannot be seen, but can be detected by its gravitational effects on other bodies.

Debris Disk
A ring-shaped circumstellar disk of dust and debris in orbit around a star. Debris disks can be created as the next phase in planetary system development following the protoplanetary disk phase. They can also be formed by collisions between planetesimals.

Declination
The angular distance of an object in the sky from the celestial equator.

Density
The amount of matter contained within a given volume. Density is measured in grams per cubic centimeter (or kilograms per liter). The density of water is 1.0, iron is 7.9, and lead is 11.3.

Disk
The surface of the Sun or other celestial body projected against the sky.

Double Asteroid
Two asteroids that revolve around each other and are held together by the gravity between them. Also called a binary asteroid.

Doppler Effect
The apparent change in wavelength of sound or light emitted by an object in relation to an observer's position. An object approaching the observer will have a shorter wavelength (blue) while an object moving away will have a longer (red) wavelength. The Doppler effect can be used to estimate an object's speed and direction.

Double Star
A grouping of two stars. This grouping can be apparent, where the stars seem close together, or physical, such as a binary system.

Dwarf Planet
A celestial body orbiting the Sun that is massive enough to be rounded by its own gravity but has not cleared its neighboring region of planetesimals and is not a satellite. It has to have sufficient mass to overcome rigid body forces and achieve hydrostatic

equilibrium. Pluto is considered to be a dwarf planet.

E

Eccentricity
The measure of how an object's orbit differs from a perfect circle. Eccentricity defines the shape of an object's orbit.

Eclipse
The total or partial blocking of one celestial body by another.

Eclipsing Binary
A binary system where one object passes in front of the other, cutting off some or all of its light.

Ecliptic
An imaginary line in the sky traced by the Sun as it moves in its yearly path through the sky.

Ejecta
Material from beneath the surface of a body such as a moon or planet that is ejected by an impact such as a meteor and distributed around the surface. Ejecta usually appear as a lighter color than the surrounding surface.

Electromagnetic Radiation
Another term for light. Light waves created by fluctuations of electric and magnetic fields in space.

Electromagnetic Spectrum
The full range of frequencies, from radio waves to gamma waves, that characterizes light.

Ellipse
An ellipse is an oval shape. Johannes Kepler discovered that the orbits of the planets were elliptical in shape rather than circular.

Elliptical Galaxy
A galaxy whose structure shaped like an ellipse and is smooth and lacks complex structures such as spiral arms.

Elongation
The angular distance of a planetary body from the Sun as seen from Earth. A planet at greatest eastern elongation is seen at its highest point above the horizon in the evening sky and a planet at greatest western elongation will be seen at its highest point above the horizon in the morning sky.

Ephemeris
A table of data arranged by date. Ephemeris tables are typically to list the positions of the Sun, Moon, planets and other solar system objects.

Equinox
The two points at which the Sun crosses the celestial equator in its yearly path in the sky. The equinoxes occur on or near March 21 and September 22. The equinoxes signal the start of the Spring and Autumn seasons.

Escape Velocity
The speed required for an object to escape the gravitational pull of a planet or other body.

Event Horizon
The invisible boundary around a black hole past which nothing can escape the gravitational pull - not even light.

Evolved Star
A star that is near the end of its life cycle where most of its fuel has been used up. At this point the star begins to loose mass in the form of stellar wind.

Extinction
The apparent dimming of star or planet when low on the horizon due to absorption by the Earth's atmosphere.

Extragalactic
A term that means outside of or beyond our own galaxy.

Extraterrestrial
A term used to describe anything that does not originate on Earth.

Eyepiece
The lens at the viewing end of a telescope. The eyepiece is responsible for enlarging the image captured by the instrument. Eyepieces are available in different powers, yielding differing amounts of magnification.

F

Faculae
Bright patches that are visible on the Sun's surface, or photosphere.

Filament
A strand of cool gas suspended over the photosphere by magnetic fields, which appears dark as seen against the disk of the Sun.

Finder
A small, wide-field telescope attached to a larger telescope. The finder is used to help point the larger telescope to the desired viewing location.

Fireball
An extremely bright meteor. Also known as bolides, fireballs can be several times brighter than the full Moon. Some can even be accompanied by a sonic boom.

Flare Star
A faint red star that appears to change in brightness due to explosions on its surface.

G

Galactic Halo
The name given to the spherical region surrounding the center, or nucleus of a galaxy.

Galactic Nucleus
A tight concentration of stars and gas found at the innermost regions of a galaxy. Astronomers now

believe that massive black holes may exist in the center of many galaxies.

Galaxy
A large grouping of stars. Galaxies are found in a variety of sizes and shapes. Our own Milky Way galaxy is spiral in shape and contains several billion stars. Some galaxies are so distant their light takes millions of years to reach the Earth.

Galilean Moons
The name given to Jupiter's four largest moons, Io, Europa, Callisto, and Ganymede. They were discovered independently by Galileo Galilei and Simon Marius.

Gamma-ray
The highest energy, shortest wavelength form of electromagnetic radiation.

Geosynchronous Orbit
An orbit in which a satellite's orbital velocity is matched to the rotational velocity of the planet. A spacecraft in geosynchronous orbit appears to hang motionless above one position of a planet's surface.

Giant Molecular Cloud (GMC)
Massive clouds of gas in interstellar space composed primarily of hydrogen molecules. These

clouds have enough mass to produce thousands of stars and are frequently the sites of new star formation.

Globular Cluster
A tight, spherical grouping of hundreds of thousands of stars. Globular clusters are composed of older stars, and are usually found around the central regions of a galaxy.

Granulation
A pattern of small cells that can be seen on the surface of the Sun. They are caused by the convective motions of the hot gases inside the Sun.

Gravitational Lens
A concentration of matter such as a galaxy or cluster of galaxies that bends light rays from a background object. Gravitational lensing results in duplicate images of distant objects.

Gravity
A mutual physical force of nature that causes two bodies to attract each other.

Greenhouse Effect
An increase in temperature caused when incoming solar radiation is passed but outgoing thermal radiation is blocked by the atmosphere. Carbon

dioxide and water vapor are two of the major gases responsible for this effect.

H

Heliopause
The point in space at which the solar wind meets the interstellar medium or solar wind from other stars.

Heliosphere
The space within the boundary of the heliopause containing the Sun and the Solar System.

Hydrogen
An element consisting of one electron and one proton. Hydrogen is the lightest of the elements and is the building block of the universe. Stars form from massive clouds of hydrogen gas.

Hubble's Law
The law of physics that states that the farther a galaxy is from us, the faster it is moving away from us.

Hydrostatic equilibrium
A state that occurs when compression due to gravity is balanced by a pressure gradient which creates a pressure gradient force in the opposite direction. Hydrostatic equilibrium is responsible for keeping

stars from imploding and for giving planets their spherical shape.

Hypergalaxy
A system consisting of a spiral galaxy surrounded by several dwarf white galaxies, often ellipticals. Our galaxy and the Andromeda galaxy are examples of hypergalaxies.

I

Ice
A term used to describe water or a number of gases such as methane or ammonia when in a solid state.

Inclination
A measure of the tilt of a planet's orbital plane in relation to that of the Earth.

Inferior Conjunction
A conjunction of an inferior planet that occurs when the planet is lined up directly between the Earth and the Sun.

Inferior Planet
A planet that orbits between the Earth and the Sun. Mercury and Venus are the only two inferior planets in our solar system.

International Astronomical Union (IAU)
An international organization that unites national astronomical societies from around the world and acts as the internationally recognized authority for assigning designations to celestial bodies and their surface features.

Interplanetary Magnetic Field
The magnetic field carried along with the solar wind.

Interstellar Medium
The gas and dust that exists in open space between the stars.

Ionosphere
A region of charged particles in a planet's upper atmosphere. In Earth's atmosphere, the ionosphere begins at an altitude of about 25 miles and extends outward about 250.

Iron Meteorite
A meteorite that is composed mainly of iron mixed with smaller amounts of nickel.

Irregular Galaxy
A galaxy with no spiral structure and no symmetric shape. Irregular galaxies are usually filamentary or very clumpy in shape.

Irregular Satellite
A satellite that orbits a planet far away with an orbit that is eccentric and inclined. They also tend to have retrograde orbits. Irregular satellites are believed to have been captured by the planet's gravity rather than being formed along with the planet.

J

Jansky
A unit used in radio astronomy to indicate the flux density (the rate of flow of radio waves) of electromagnetic radiation received from outer space. A typical radio source has a spectral flux density of roughly 1 Jy. The jansky was named to honor Karl Gothe Jansky who developed radio astronomy in 1932.

Jet
A narrow stream of gas or particles ejected from an accretion disk surrounding a star or black hole.

K

Kelvin
A temperature scale used in sciences such as astronomy to measure extremely cold temperatures. The Kelvin temperature scale is just like the Celsius

scale except that the freezing point of water, zero degrees Celsius, is equal to 273 degrees Kelvin. Absolute zero, the coldest known temperature, is reached at 0 degrees Kelvin or -273.16 degrees Celsius.

Kepler's First Law
A planet orbits the Sun in an ellipse with the Sun at one focus.

Kepler's Second Law
A ray directed from the Sun to a planet sweeps out equal areas in equal times.

Kepler's Third Law
The square of the period of a planet's orbit is proportional to the cube of that planet's semi major axis; the constant of proportionality is the same for all planets.

Kiloparsec
A distance equal to 1000 parsecs.

Kirkwood Gaps
Regions in the main belt of asteroids where few or no asteroids are found. They were named after the scientist who first noticed them.

Kuiper Belt
A large ring of icy, primitive objects beyond the orbit of Neptune. Kuiper Belt objects are believed to be remnants of the original material that formed the Solar System. Some astronomers believe Pluto and Charon are Kuiper Belt objects.

L

Lagrange Point
French mathematician and astronomer Joseph Louis Lagrange showed that three bodies could lie at the apexes of an equilateral triangle which rotates in its plane. If one of the bodies is sufficiently massive compared with the other two, then the triangular configuration is apparently stable. Such bodies are sometimes referred to as Trojans. The leading apex of the triangle is known as the leading Lagrange point or L4; the trailing apex is the trailing Lagrange point or L5.

Lenticular Galaxy
A disk-shaped galaxy that contains no conspicuous structure within the disk. Lenticular galaxies tend to look more like elliptical galaxies than spiral galaxies.

Libration
An effect caused by the apparent wobble of the Moon as it orbits the Earth. The Moon always keeps the same side toward the Earth, but due to libration, 59% of the Moon's surface can be seen over a period of time.

Light Year
An astronomical unit of measure equal to the distance light travels in a year, approximately 5.8 trillion miles.

Limb
The outer edge or border of a planet or other celestial body.

Local Group
A small group of about two dozen galaxies of which our own Milky Way galaxy is a member.

Luminosity
The amount of light emitted by a star.

Lunar Eclipse
A phenomenon that occurs when the Moon passes into the shadow of the Earth. A partial lunar eclipse occurs when the Moon passes into the penumbra, or partial shadow. In a total lunar eclipse, the Moon passes into the Earth's umbra, or total shadow.

Lunar Month
The average time between successive new or full moons. A lunar month is equal to 29 days 12 hours 44 minutes. Also called a synodic month.

Lunation
The interval of a complete lunar cycle, between one new Moon and the next. A lunation is equal to 29 days, 12 hours, and 44 minutes.

M

Magellanic Clouds
Two small, irregular galaxies found just outside our own Milky Way galaxy. The Magellanic Clouds are visible in the skies of the southern hemisphere.

Magnetic Field
A condition found in the region around a magnet or an electric current, characterized by the existence of a detectable magnetic force at every point in the region and by the existence of magnetic poles.

Magnetic Pole
Either of two limited regions in a magnet at which the magnet's field is most intense.

Magnetosphere
The area around a planet most affected by its magnetic field. The boundary of this field is set by the solar wind.

Magnitude
The degree of brightness of a star or other object in the sky according to a scale on which the brightest star has a magnitude -1.4 and the faintest visible star has magnitude 6. Sometimes referred to as apparent magnitude. In this scale, each number is 2.5 times the brightness of the previous number. Thus a star with a magnitude of 1 is 100 times brighter than on with a visual magnitude of 6.

Main Belt
The area between Mars and Jupiter where most of the asteroids in our solar system are found.

Major Planet
A name used to describe any planet that is considerably larger and more massive than the Earth, and contains large quantities of hydrogen and helium. Jupiter and Neptune are examples of major planets.

Mare
A term used to describe a large, circular plain. The word mare means "sea". On the Moon, the maria are the smooth, dark-colored areas.

Mass
A measure of the total amount of material in a body, defined either by the inertial properties of the body or by its gravitational influence on other bodies.

Matter
A word used to describe anything that contains mass.

Meridian
An imaginary circle drawn through the North and South poles of the celestial equator.

Metal
A term used by astronomers to describe all elements except hydrogen and helium, as in "the universe is composed of hydrogen, helium and traces of metals". This astronomical definition is quite different from the traditional chemistry definition of a metal.

Meteor
A small particle of rock or dust that burns away in the Earth's atmosphere. Meteors are also referred to as shooting stars.

Meteor Shower
An event where a large number of meteors enter the Earth's atmosphere from the same direction in space at nearly the same time. Most meteor showers take place when the Earth passes through the debris left behind by a comet.

Meteorite
An object, usually a chunk or metal or rock, that survives entry through the atmosphere to reach the Earth's surface. Meteors become meteorites if they reach the ground.

Meteoroid
A small, rocky object in orbit around the Sun, smaller than an asteroid.

Millibar
A measure of atmospheric pressure equal to 1/1000 of a bar. Standard sea-level pressure on Earth is about 1013 millibars.

Minor Planet
A term used since the 19th century to describe objects, such as asteroids, that are in orbit around the Sun but are not planets or comets. In 2006, the International Astronomical Union reclassified minor planets as either dwarf planets or small solar system bodies.

Molecular Cloud
An interstellar cloud of molecular hydrogen containing trace amounts of other molecules such as carbon monoxide and ammonia.

N

Nadir
A term used to describe a point directly underneath an object or body.

Nebula
A cloud of dust and gas in space, usually illuminated by one or more stars. Nebulae represent the raw material the stars are made of.

Neutrino
A fundamental particle produced by the nuclear reactions in stars. Neutrinos are very hard to detect because the vast majority of them pass completely through the Earth without interacting.

Neutron Star
A compressed core of an exploded star made up almost entirely of neutrons. Neutron stars have a strong gravitational field and some emit pulses of energy along their axis. These are known as pulsars.

Newton's First Law of Motion
A body continues in its state of constant velocity (which may be zero) unless it is acted upon by an external force.

Newton's Second Law of Motion
For an unbalanced force acting on a body, the acceleration produced is proportional to the force impressed; the constant of proportionality is the inertial mass of the body.

Newton's Third Law of Motion
In a system where no external forces are present, every action force is always opposed by an equal and opposite reaction.

Nova
A star that flares up to several times its original brightness for some time before returning to its original state.

Nuclear Fusion
The nuclear process whereby several small nuclei are combined to make a larger one whose mass is slightly smaller than the sum of the small ones. Nuclear fusion is the reaction that fuels the Sun, where hydrogen nuclei are fused to form helium.

O

Obliquity
The angle between a body's equatorial plane and orbital plane.

Oblateness
A measure of flattening at the poles of a planet or other celestial body.

Occultation
An event that occurs when one celestial body conceals or obscures another. For example, a solar eclipse is an occultation of the Sun by the Moon.

Oort Cloud
A theoretical shell of comets that is believed to exist at the outermost regions of our solar system. The Oort cloud was named after the Dutch astronomer who first proposed it.

Open Cluster

A collection of young stars that formed together. They may or may not be still bound by gravity. Some of the youngest open clusters are still embedded in the gas and dust from which they formed.

Opposition

The position of a planet when it is exactly opposite the Sun in the sky as seen from Earth. A planet at opposition is at its closest approach to the Earth and is best suitable for observing.

Orbit

The path of a celestial body as it moves through space.

P

Parallax

The apparent change in position of two objects viewed from different locations.

Parsec

A large distance often used in astronomy. A parsec is equal to 3.26 light-years.

Patera

A shallow crater with a complex, scalloped edge.

Penumbra
The area of partial illumination surrounding the darkest part of a shadow caused by an eclipse.

Perigee
The point in the orbit of the Moon or other satellite at which it is closest to the Earth.

Perihelion
The point in the orbit of a planet or other body where it is closest to the Sun.

Perturb
To cause a planet or satellite to deviate from a theoretically regular orbital motion.

Phase
The apparent change in shape of the Moon and inferior planets as seen from Earth as they move in their orbits.

Photon
A particle of light composed of a minute quantity of electromagnetic energy.

Photosphere
The bright visible surface of the Sun.

Planemo
A large planet or planetary body that does not orbit a star. Planemos instead wander cold and alone through the cosmos. It is believed that most planemos once orbited their mother star but were ejected from the star system by gravitational interaction with another massive object.

Planet
A celestial body orbiting a star or stellar remnant that is massive enough to be rounded by its own gravity, is not massive enough to cause thermonuclear fusion, and has cleared its neighboring region of planetesimals.

Planetary Nebula
A shell of gas surrounding a small, white star. The gas is usually illuminated by the star, producing a variety of colors and shapes.

Planetesimal
A solid object that is believed to exist in protoplanetary disks and in debris disks. Planetesimals are formed from small dust grains that collide and stick together and are the building blocks that eventually form planets in new planetary systems.

Planitia
A low plain.

Planum
A high plain or plateau.

Plasma
A form of ionized gas in which the temperature is too high for atoms to exist in their natural state. Plasma is composed of free electrons and free atomic nuclei.

Precession
The apparent shift of the celestial poles caused by a gradual wobble of the Earth's axis.

Prominence
An explosion of hot gas that erupts from the Sun's surface. Solar prominences are usually associated with sunspot activity and can cause interference with communications on Earth due to their electromagnetic effects on the atmosphere.

Prograde Orbit
In reference to a satellite, a prograde orbit means that the satellite orbits the planet in the same direction as the planet's rotation. A planet is said to have a prograde orbit if the direction of its orbit is

the same as that of the majority of other planets in the system.

Proper Motion
The apparent angular motion across the sky of an object relative to the Solar System.

Protoplanetary Disk
A rotating circumstellar disk of dense gas surrounding a young newly formed star. It is thought that planets are eventually formed from the gas and dust within the protoplanetary disk.

Protostar
Dense regions of molecular clouds where stars are forming.

Pulsar
A spinning neutron star that emits energy along its gravitational axis. This energy is received as pulses as the star rotates.

Q

Quadrature
A point in the orbit of a superior planet where it appears at right angles to the Sun as seem from Earth.

Quasar
An unusually bright object found in the remote areas of the universe. Quasars release incredible amounts of energy and are among the oldest and farthest objects in the known universe. They may be the nuclei of ancient, active galaxies.

Quasi-Stellar Object
Sometimes also called quasi-stellar source, this is a star-like object with a large redshift that gives off a strong source of radio waves. They are highly luminous and presumed to be extragalactic.

R

Radial Velocity
The movement of an object either towards or away from a stationary observer.

Radiant
A point in the sky from which meteors in a meteor shower seem to originate.

Radiation
Energy radiated from an object in the form of waves or particles.

Radiation Belt
Regions of charged particles in a magnetosphere.

Radio Galaxy
A galaxy that gives off large amounts of energy in the form of radio waves.

Red Giant
A stage in the evolution of a star when the fuel begins to exhaust and the star expands to about fifty times its normal size. The temperature cools, which gives the star a reddish appearance.

Redshift
A shift in the lines of an object's spectrum toward the red end. Redshift indicates that an object is moving away from the observer. The larger the redshift, the faster the object is moving.

Regular Satellite
A satellite that orbits close to a planet in a nearly circular, equatorial orbit. Regular satellites are believed to have been formed at the same time as the planet, unlike irregular satellites which are believed to have been captured by the planet's gravity.

Resonance
A state in which an orbiting object is subject to periodic gravitational perturbations by another.

Retrograde Motion
The phenomenon where a celestial body appears to slow down, stop, then move in the opposite direction. This motion is caused when the Earth overtakes the body in its orbit.

Retrograde Orbit
The orbit of a satellite where the satellite travels in a direction opposite to that direction of the planet's rotation.

Right Ascension
The amount of time that passes between the rising of Aries and another celestial object. Right ascension is one unit of measure for locating an object in the sky.

Ring Galaxy
A galaxy that has a ring-like appearance. The ring usually contains luminous blue stars. Ring galaxies are believed to have been formed by collisions with other galaxies.

Roche Limit
The smallest distance from a planet or other body at which purely gravitational forces can hold together a satellite or secondary body of the same mean density as the primary. At a lesser distance the tidal forces of the primary would break up the secondary.

Rotation
The spin of a body about its axis.

S

Saber's Beads
Broken arc of illuminations seen at the limb of very young or old lunar crescents. The visual similarity to the moments before and after a total solar eclipse was first noted by American astronomer Stephen Saber.

Saros Series
Also known as a saros cycle, a period of 223 synodic months that can be used to predict solar and lunar eclipses. The saros cycle is equal to 6,585.3 days (18 years 11 days 8 hours).

Satellite
A natural or artificial body in orbit around a planet.

Scarp
A line of cliffs produced erosion or by the action of faults.

Seyfert Galaxy
A type of spiral galaxy which has a small, compact nucleus that is much brighter than the rest of the

galaxy. The nucleus exhibits variable light intensity and radio emission suggesting that a black hole may be devouring material at the galaxy's center.

Shell Star
A type of star which is believed to be surrounded by a thin envelope of gas, which is often indicated by bright emission lines in its spectrum.

Shepherd Satellite
A satellite that constrains the extent of a planetary ring through gravitational forces. Also known as a shepherd moon.

Sidereal
Of, relating to, or concerned with the stars. Sidereal rotation is that measured with respect to the stars rather than with respect to the Sun or the primary of a satellite.

Sidereal Month
The average period of revolution of the Moon around the Earth in reference to a fixed star, equal to 27 days, 7 hours, 43 minutes in units of mean solar time.

Sidereal Period
The period of revolution of a planet around the Sun or a satellite around its primary.

Singularity
The center of a black hole, where the curvature of space time is maximal. At the singularity, the gravitational tides diverge. Theoretically, no solid object can survive hitting the singularity.

Small Solar System Body
A term defined in 2006 by the International Astronomical Union to describe objects in the Solar System that are neither planets nor dwarf planets. These include most of the asteroids, comets, and other small bodies in the Solar System.

Solar Cycle
The approximately 11-year quasi-periodic variation in frequency or number of solar active events.

Solar Eclipse
A phenomenon that occurs when the Earth passes into the shadow of the Moon. A total solar eclipse occurs when the Moon is close enough to completely block the Sun's light. An annular solar eclipse occurs when the Moon is farther away and is not able to completely block the light. This results in a ring of light around the Moon.

Solar Flare
A bright eruption of hot gas in the Sun's photosphere. Solar prominences are usually only detectable by specialized instruments but can be visible during a total solar eclipse.

Solar Nebula
The cloud of dust and gas out of which the Solar System was believed to have formed about 5 billion years ago.

Solar Wind
A flow of charged particles that travels from the Sun out into the Solar System.

Solstice
The time of the year when the Sun appears furthest north or south of the celestial equator. The solstices mark the beginning of the Summer and Winter seasons.

Spectrometer
The instrument connected to a telescope that separates the light signals into different frequencies, producing a spectrum.

Spectroscopy

The technique of observing the spectra of visible light from an object to determine its composition, temperature, density, and speed.

Spectrum
The range of colors that make up visible white light. A spectrum is produced when visible light passes through a prism.

Spicules
Grass-like patterns of gas seen in the atmosphere of the Sun.

Spiral Galaxy
A galaxy that contains a prominent central bulge and luminous arms of gas, dust, and young stars that wind out from the central nucleus in a spiral formation. Our galaxy, the Milky Way, is a spiral galaxy.

Star
A giant ball of hot gas that creates and emits its own radiation through nuclear fusion.

Star Cluster
A large grouping of stars, from a few dozen to a few hundred thousand, that are bound together by their mutual gravitational attraction.

Steady State Theory
The theory that suggests the universe is expanding but exists in a constant, unchanging state in the large scale. The theory states that new matter is being continually being created to fill the gaps left by expansion. This theory has been abandoned by most astronomers in favor of the big bang theory.

Stellar Wind
The ejection of gas from the surface of a star. Many different types of stars, including our Sun, have stellar winds. The stellar wind of our Sun is also known as the Solar wind. A star's stellar wind is strongest near the end of its life when it has consumed most of its fuel.

Stone Meteorite
A meteorite that resembles a terrestrial rock and is composed of similar materials.

Stony Iron
A meteorite that contains regions resembling both a stone meteorite and an iron meteorite.

Sunspot
Areas of the Sun's surface that are cooler than surrounding areas. The usually appear black on visible light photographs of the Sun. Sunspots are

usually associated disturbances in the Sun's electromagnetic field.

Supergiant
The stage in a star's evolution where the core contracts and the star swells to about five hundred times its original size. The star's temperature drops, giving it a red color.

Supermoon
A term used to describe a full moon that occurs during the Moon's closest approach to the Earth. During a supermoon, the Moon may appear slightly larger and brighter than normal.

Superior Conjunction
A conjunction that occurs when a planet passes behind the Sun and is on the opposite side of the Sun from the Earth.

Superior Planet
A planet that exists outside the orbit of the Earth. All of the planets in our solar system are superior except for Mercury and Venus. These two planets are inferior planets.

Supernova
A supernova is a cataclysmic explosion caused when a star exhausts its fuel and ends its life.

Supernovae are the most powerful forces in the universe. All of the heavy elements were created in supernova explosions.

Supernova Remnant
An expanding shell of gas ejected at high speeds by a supernova explosion. Supernova remnants are often visible as diffuse gaseous nebulae usually with a shell-like structure. Many resemble "bubbles" in space.

Synchronous Rotation
A period of rotation of a satellite about its axis that is the same as the period of its orbit around its primary. This causes the satellite to always keep the same face to the primary. Our Moon is in synchronous rotation around the Earth.

Synodic Month
The period of time it takes the Moon to make one complete revolution around the Earth. A Synodic month is equal to 29.53 days and is measured as the time between a lunar phase and the return of that same phase.

Synodic Period
The interval between points of opposition of a superior planet.

T

Tektite
A small, glassy material formed by the impact of a large body, usually a meteor or asteroid. Tektites are commonly found at the sites of meteor craters.

Telescope
An instrument that uses lenses and sometimes mirrors to collect large amounts of light from distant objects and enable direct observation and photography. A Telescope can also include any instrument designed to observe distant objects by their emissions of invisible radiation such as x-rays or radio waves.

Terminator
The boundary between the light side and the dark side of a planet or other body.

Terrestrial
A term used to describe anything originating on the planet Earth.

Terrestrial Planet
A name given to a planet composed mainly of rock and iron, similar to that of Earth.

Tidal Force
The differential gravitational pull exerted on any extended body within the gravitational field of another body.

Tidal Heating
Frictional heating of a satellite's interior due to flexure caused by the gravitational pull of its parent planet and/or other neighboring satellites.

Transit
The passage of a celestial body across an observer's meridian; also the passage of a celestial body across the disk of a larger one.

Trans-Neptunian Object (TNO)
Any one of a number of celestial objects that orbit the Sun at a distance beyond the orbit of the planet Neptune.

Trojan
An object orbiting in the Lagrange points of another (larger) object. This name derives from a generalization of the names of some of the largest asteroids in Jupiter's Lagrange points. Saturn's moons Helene, Calypso and Telesto are also sometimes called Trojans.

U

Ultraviolet
Electromagnetic radiation at wavelengths shorter than the violet end of visible light. The atmosphere of the Earth effectively blocks the transmission of most ultraviolet light, which can be deadly to many forms of life.

Umbra
The area of total darkness in the shadow caused by an eclipse.

Universal Time (UT)
Also known as Greenwich Mean Time, this is local time on the Greenwich meridian. Universal time is used by astronomers as a standard measure of time.

V

Van Allen Belts
Radiation zones of charged particles that surround the Earth. The shape of the Van Allen belts is determined by the Earth's magnetic field.

Variable Star
A star that fluctuates in brightness. These include eclipsing binaries.

Visible Light
Wavelengths of electromagnetic radiation that are visible to the human eye.

Virgo Cluster
A gigantic cluster of over 2000 galaxies that is located mainly within the constellation of Virgo. This cluster is located about 60 million light-years from Earth.

Visual Magnitude
A scale used by astronomers to measure the brightness of a star or other celestial object. Visual magnitude measures only the visible light from the object. On this scale, bright objects have a lower number than dim objects.

W

Wavelength
The distance between consecutive crests of a wave. This serves as a unit of measure of electromagnetic radiation.

White Dwarf
A very small, white star formed when an average sized star uses up its fuel supply and collapses. This process often produces a planetary nebula, with the white dwarf star at its center.

X

X-ray
Electromagnetic radiation of a very short wavelength and very high-energy. X-rays have shorter wavelengths than ultraviolet light but longer wavelengths than cosmic rays.

X-ray Astronomy
The field of astronomy that studies celestial objects by the x-rays they emit.

X-ray Star
A bright celestial object that gives off x-rays as a major portion of its radiation.

Y

Yellow Dwarf
An ordinary star such as the Sun at a stable point in its evolution.

Z

Zenith
A point directly overhead from an observer.

Zodiac

An imaginary belt across the sky in which the Sun, moon, and all of the planets can always be found.

Zodiacal Light
A faint cone of light that can sometimes be seen above the horizon after sunset or before sunrise. Zodiacal light is caused by sunlight reflecting off small particles of material in the plane of the Solar System.